Machines with Power!

Garbage Trucks

by Amy McDonald

BELLWETHER MEDIA
MINNEAPOLIS, MN

Blastoff! Beginners are developed by literacy experts and educators to meet the needs of early readers. These engaging informational texts support young children as they begin reading about their world. Through simple language and high frequency words paired with crisp, colorful photos, Blastoff! Beginners launch young readers into the universe of independent reading.

Sight Words in This Book

a	down	the
are	for	they
at	here	this
big	is	to
come	it	up

This edition first published in 2022 by Bellwether Media, Inc.

Library of Congress Cataloging-in-Publication Data

Names: McDonald, Amy, author.
Title: Garbage trucks / by Amy McDonald.
Description: Minneapolis, MN : Bellwether Media, Inc., 2022. | Series: Blastoff! Beginners : Machines with power! | Includes bibliographical references and index. | Audience: Ages 4-7 | Audience: Grades K-1
Identifiers: LCCN 2021003770 (print) | LCCN 2021003771 (ebook) | ISBN 9781644874776 (library binding) | ISBN 9781648343858 (ebook)
Subjects: LCSH: Refuse collection vehicles--Juvenile literature.
Classification: LCC TD792 .M365 2021 (print) | LCC TD792 (ebook) | DDC 628.4/420284--dc23
LC record available at https://lccn.loc.gov/2021003770
LC ebook record available at https://lccn.loc.gov/2021003771

Editor: Christina Leaf Designer: Andrea Schneider

Printed in the United States of America, North Mankato, MN.

Table of Contents

What Are Garbage Trucks?

Stand back! Here comes a garbage truck!

ALLIED WASTE SERVICES
2156
MACK

Garbage trucks
are big machines.
They pick up trash.

trash

Parts of a Garbage Truck

This is the arm.
It picks up bins.

arm
bins

This is the
hopper.
It holds trash.

hopper
195
McNeilus

This is the **blade**.
It pushes
trash down.

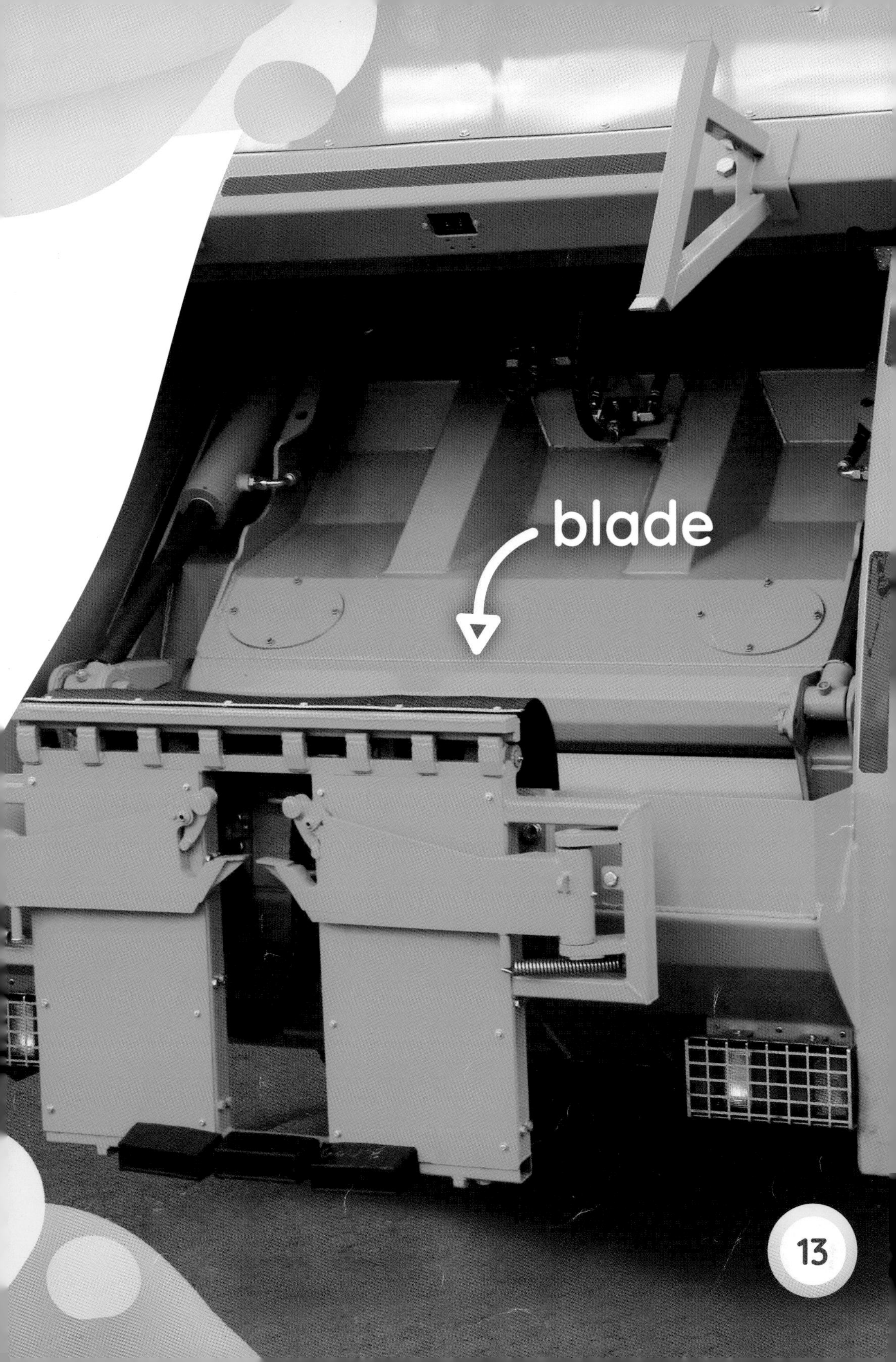
blade

This is the **cab**.
It is for the driver.

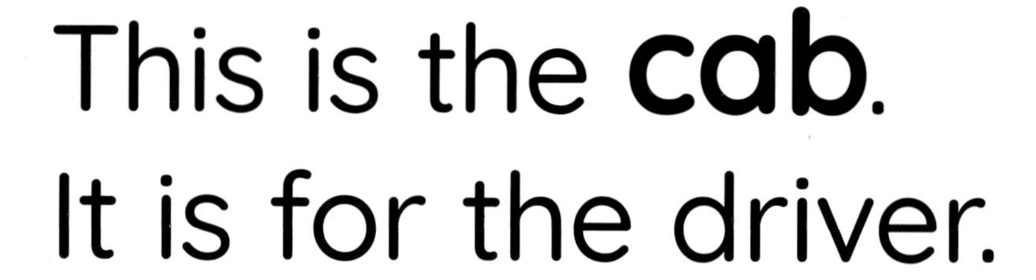

5576
8
WM
WASTE MANAGEMENT
Think Green.
CNG
104424
CA 39065

Garbage Trucks at Work

This truck holds a **dumpster**. It is at a job site.

dumpster

This truck
lifts dumpsters.
It is strong!

McNeilus
325
DORADO
SERVICES, INC
757-398-9090
USDOT 1424969VA

This truck goes
to homes.
Here it comes!

405

Garbage Truck Facts

Garbage Truck Parts

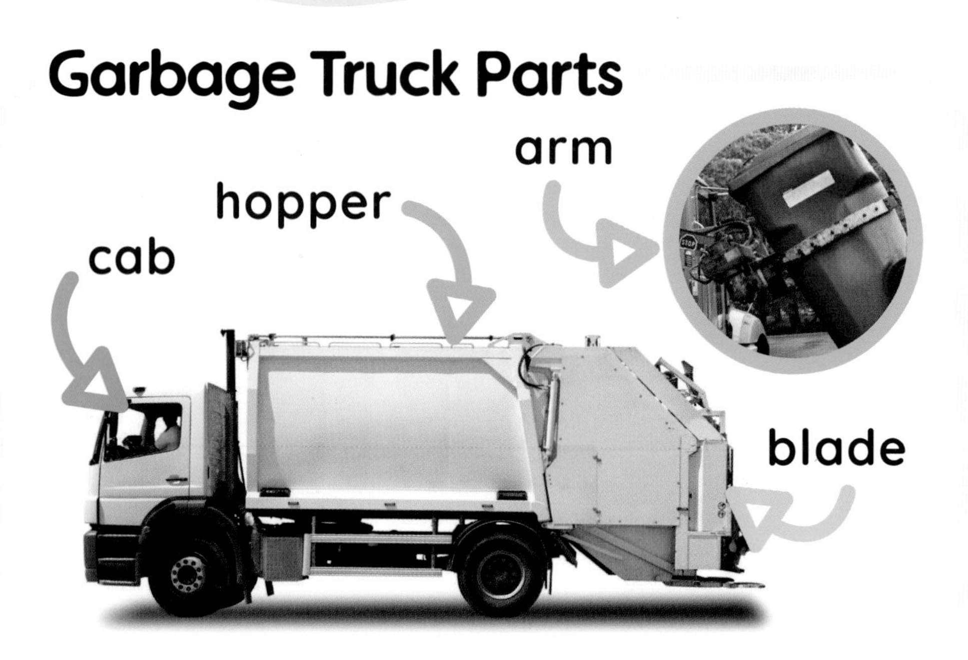

Garbage Truck Stops

job sites

offices

homes

Glossary

blade

a tool that flattens trash

cab

a place where the driver sits

dumpster

a large trash bin

hopper

the box that holds trash on a truck

To Learn More

ON THE WEB

FACTSURFER

Factsurfer.com gives you a safe, fun way to find more information.

1. Go to www.factsurfer.com.
2. Enter "garbage trucks" into the search box and click 🔍.
3. Select your book cover to see a list of related content.

Index

The images in this book are reproduced through the courtesy of: Art Konovalov, front cover; Konstantinos Moraitis, p. 3; Michael T Hartman, pp. 4-5; DeawSS, p. 6; Nadya So, pp. 6-7; Carolyn Franks, pp. 8-9; Polina MB, pp. 10-11; ilmarinfoto, pp. 12-13; Robwilson39, pp. 14, 22 (garbage truck parts); Bill Morson, pp. 14-15, 23 (cab); Johan Larson, p. 16; ungvar, pp. 16-17, 22 (job sites); PJF Military Collection/ Alamy, pp. 18-19; Sunrise Photos/ Alamy, pp. 20-21; Paul Vasahelyi, p. 22 (arm); Dmitry Kalinovsky, p. 22 (offices); Roman Yanushevsky, p. 22 (homes); People Images, p. 23 (blade); StudioPortoSabbia, p. 23 (dumpster); Africa Studio, p. 23 (hopper).